Machir

Joy Brewster

Contents

Machines Help You Work

Picture yourself pushing a friend on a swing or throwing a ball. Both sound like fun, right? They're also **work.** In science, work has a very specific meaning. You do work any time you move something—even yourself.

Work has two parts. It includes a **force**—something that sets the object in motion. Work also includes the **distance** that you move an object. For example, if you ride a bike to school, the distance is the number of miles that you travel.

It takes an awful lot of work to push a car across a field. In this case, pushing is the force that's used and how far they move the car is the distance. Pushing, pulling, lifting, squeezing, and twisting are different kinds of force.

It's not always easy to do work by yourself. There are times when you need the help of a machine. Machines can decrease the amount of force that you need to apply, so you don't have to push or pull as hard.

Which of these tools is a machine? They both are. The backhoe is very complex. It has many moving parts and is powered by an engine. The shovel, however, is a **simple machine.** It has few or no moving parts and is people-powered.

Simple Machines

There are six simple machines that help you work. They change the direction of the force or change the distance to make work easier.

The Inclined Plane

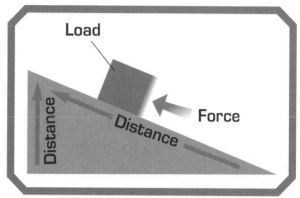

Suppose you need to move a heavy box from a sidewalk up to a porch. How would you do it? Pushing it up a simple machine called an **inclined plane** would be easier than lifting it. Your box would move across a longer distance, but you wouldn't have to use as much muscle. An inclined plane is a flat surface that is slanted up or down. Inclined planes make work easier by changing the distance over which the work is done.

Inclined planes can also help you lower objects. These blocks of ice are heavy. It would take two workers to lift one down to the ground. Using this simple machine, though, a single worker can slowly push an ice block down a ramp.

The Wedge

Anytime you cut food with a knife, you're using a simple machine called a **wedge.** Wedges help us in three ways: they cut objects into pieces, push objects apart, or hold them together. An ax, a spatula, and a doorstop are all examples of wedges. They make work easier by changing the direction of the force. For instance, when the blade of an ax strikes a piece of wood, the downward movement becomes a force that pushes out from the sides and cuts the wood.

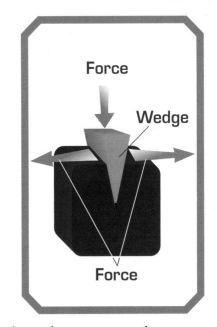

Force

Wedge

Force

The downward, slanting movement of this chisel changes to a force that pushes upward at an angle. This strong force can easily chip off pieces from a block of wood.

Quick Fact!

When you bite into a piece of food, your front teeth act as a wedge. They separate the food into pieces that you can easily chew.

The Screw

The **screw,** another simple machine, can fasten objects together, bore holes, or lift things. A screw has a raised edge called a **thread.** If you could unwind the thread it would be much longer than the length of the screw. As a screw turns round and round, it travels along a spiral path that is cut by this thread. So, the screw travels a greater distance than if it were pushed straight in, which makes the work easier. Inclined planes, wedges, and screws all belong to one family of simple machines.

Rotating force

Thread

Pulling force

Screws change the distance and also the direction of the force. As a screw turns, the rotating force changes direction to pull the screw forward. This strong force can be used to raise objects or fasten them together.

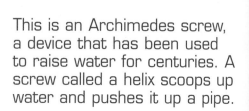

This is an Archimedes screw, a device that has been used to raise water for centuries. A screw called a helix scoops up water and pushes it up a pipe.

The Wheel and Axle

Whenever you open a door, you use a simple machine called a **wheel and axle.** A doorknob is attached to a thin rod—an axle. You turn the knob—or wheel—to turn the axle and open the door. Without the wheel, it would take a lot of force to turn the axle using just your fingers. Like some of the other simple machines, a wheel and axle helps us do work by changing the distance.

As the wheel and axle on this faucet turn together, the wheel spins in a larger circle and travels a greater distance. With this change in distance, you can apply a small amount of force to the wheel to turn the attached axle.

The big wheel at the bottom of the axle turns in a larger circle than the small wheel at the top. So, this pottery maker can apply a little bit of force to the bottom wheel to spin the top wheel more easily.

The Lever

It would be almost impossible to lift a grown-up by yourself. But, using a seesaw, you could do it pretty easily. A seesaw is a simple machine called a **lever.**

A lever is a bar that turns on a fixed point, or **fulcrum.** The object that you lift or turn is called the **load.** To apply force, you push or pull at some point along the lever. Levers come in three classes. In each, the fulcrum, load, and force are arranged differently.

A seesaw is a first-class lever. On this kind of lever, the fulcrum sits somewhere between the load and the force. As you push down on one end, the load is lifted up on the other. If a heavy load is close to the fulcrum, it becomes very easy to lift.

Load

Force

Fulcrum

To help you do work, first- and second-class levers change the distance, so you don't have to use as much force. With third-class levers, you need to use a lot of force, but your range of motion increases. Levers can be used to lift, turn, or propel objects, and also to pry objects loose.

This is a second-class lever. In this kind of lever, the fulcrum and the force are at either end, with the load in between. This makes it easy to lift or lower a heavy load, such as the contents of a wheelbarrow.

A tennis racket is a third-class lever. This athlete's arm is the fulcrum. The force is applied where his hand grips the handle. The part that connects with the load—the racket head—moves with a wide range of motion. Which means that it can swing far and fast to hit the ball across a great distance.

Force   Load

Fulcrum

Force   Fulcrum

Load

Investigate Levers

You need a heavy hardcover book, such as a textbook or dictionary, a ruler, and a rubber band.

1 Set a ruler on a flat table or desk with two centimeters sticking out over the edge. Place the book on the opposite end of the ruler. The book should be resting sideways, right up against the far end of the ruler. Make sure that it doesn't hang over the end. Wrap a rubber band around the book and ruler to hold them together.

2 The ruler is now a lever. You will apply force to the end that sticks out. *Identify* the fulcrum and the load. What class of lever is this? How far is the force from the fulcrum? How far is the load from the fulcrum? *Record* all of this data.

3 Now, try to raise the book by pressing down on the end of the ruler. Can you lift the book?

4 Slide the ruler toward you until three centimeters stick out. Press down on the ruler again to try to raise the book. Keep sliding the ruler toward you, one centimeter at a time, until you can raise the book. *Record* the measurement at which this is possible.

5 How many centimeters is the book from the table's edge? Has the distance between the load and fulcrum increased or decreased? How many centimeters stick out over the table? Has the distance between the force and the fulcrum increased or decreased? What can you *conclude* about levers, distance, and force?

The Pulley

A **pulley** makes it easy to lift loads to high places. This simple machine is a wheel with a rope looped around it, which spins on an axle. If you pull down on one end of the rope, you can lift a load attached to the other end. Wheels and axles, levers, and pulleys all belong to the same family of simple machines. With wheels and pulleys, the axle is the fulcrum.

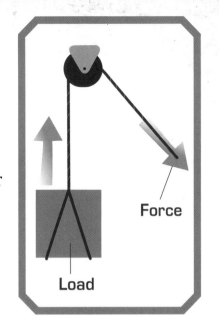

Force

Load

Pulleys change the direction of the force. So, instead of lifting up a load, you pull down on a rope, which is easier. Using compound—two or more—pulleys changes the distance, too. A rope looped over and around several different pulleys travels a greater distance than one looped around a single pulley. This makes lifting heavy loads even easier.

Compound Machines

You use simple machines every day without even thinking about it. For instance, a bottle opener is a lever and the top of a soda bottle is a screw. Most machines, however, are **compound machines.** They're made up of two or more simple machines working together. Think about a blender, for example. The blades inside are wedges, but they turn on a wheel and axle. Here are several compound machines. Can you think of others? Check the inside back cover for answers to the following questions.

Tongs combine two levers working together. The force is in between the fulcrum and the load, because you grab hold of the middle of a pair of tongs to pick up food. Knowing that, what class of lever are tongs?

A hand-powered drill combines a wheel and axle and one other simple machine. What is it?

Quick Fact!

You can even find simple machines inside electronic devices, such as computers. A disk drive uses a wheel and axle. It also has a system of levers to position the heads that read and write data.

This hand truck combines a second-class lever and a wedge with one other simple machine. What is it?

An eggbeater combines a wheel and axle with another simple machine. Can you guess what it is?

To open a can, first you pull apart the handles of an opener—like this one—then fasten it onto the can. Next, you pierce the top of the can and turn the opener all the way around it to cut open the lid. Can you guess the three simple machines that make up a can opener?

13

Gears

Large compound machines not only combine simple machines, but they can also have lots of other parts. Many machines have **gears.** Gears are wheels with teeth around the outer rim. To do work, the teeth of two gears must interlock. Then as one turns, it makes the other rotate in a different direction.

Gears come in different shapes and sizes. Notice the shape of these gears. They're used to connect parts that come together at an angle, such as the shafts in an automobile transmission.

Gears are also used to connect rotating parts, such as those in a watch. As gears rotate in different directions, they transfer force and movement from one part of a machine to another. Using gears of different sizes can help make the force stronger, too.

Chains and Belts

A chain is often used with a pulley, in place of a rope. Chains, and also belts, can be used with wheels and gears, too. They can be arranged in different ways to change the direction of the wheels. When a crossed chain or belt is used, two wheels turn in *opposite* directions. When an uncrossed chain or belt is used, they turn in the *same* direction. Like gears, chains and belts transfer force and movement from one machine part to another.

Bulldozers roll along belts attached to the wheels, which helps them move over difficult terrain. Besides the wheel and axle, can you find another simple machine on this bulldozer?

In factories, wide conveyor belts connect different machines on an assembly line. This belt is looped around giant gears. It keeps these products moving from worker to worker.

Counterweights

Some of the parts found on a machine called a crane include a lever, pulleys, and **counterweights.** They all work together to raise heavy loads. Counterweights help to keep the crane from tipping over. The weight of the load determines the amount of counterweights used.

Counterweights are also used on elevators. A cable attached to an elevator car loops around pulleys. Counterweights attached to the other end of the cable help balance the weight of the car and its passengers. As the counterweights go down, the elevator moves up, and vice versa.

Counterweights underneath this crane on the right help support the load and keep the machine from tipping forward. Counterweights are extremely heavy and can only be used during lifts. They must be removed before the crane can be driven away.

Power and Machines

With simple machines, people supply the power. We use our muscles to hammer a nail, row a boat, or shovel snow. Many compound machines, however, are powered using electricity or fuel, such as gasoline. Electric motors and engines supply the force that makes compound machines run, helping us do work. Wind, water, and sunlight can also be harnessed and used to power machines.

An electric motor drives pulleys to move an elevator car up or down. When the motor turns one way, pulleys raise the elevator. When the motor turns the other way, pulleys lower the elevator. Counterweights help support the weight of the car, so only a small amount of force is needed to raise or lower it.

This compound machine runs on gasoline or diesel fuel. Inside, an engine converts the fuel into a force that runs gears, belts, and other parts to make the tractor trailer move.

Inventing New Machines

Check out these original inventions by kids. Their designs won awards in the Young Inventors program, a science competition for students in grades 2 through 8. What simple machines do they use? You can find the answers on the inside back cover.

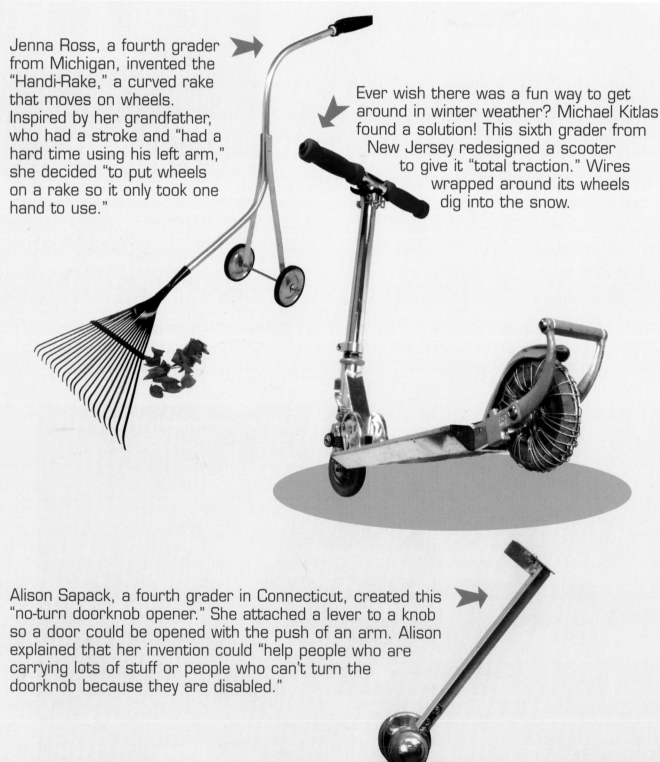

Jenna Ross, a fourth grader from Michigan, invented the "Handi-Rake," a curved rake that moves on wheels. Inspired by her grandfather, who had a stroke and "had a hard time using his left arm," she decided "to put wheels on a rake so it only took one hand to use."

Ever wish there was a fun way to get around in winter weather? Michael Kitlas found a solution! This sixth grader from New Jersey redesigned a scooter to give it "total traction." Wires wrapped around its wheels dig into the snow.

Alison Sapack, a fourth grader in Connecticut, created this "no-turn doorknob opener." She attached a lever to a knob so a door could be opened with the push of an arm. Alison explained that her invention could "help people who are carrying lots of stuff or people who can't turn the doorknob because they are disabled."

The Segway Human Transporter is a new way to get people moving. Inventor Dean Kamen originally designed a device to help people in wheelchairs. It was called the IBOT and could raise its passenger to a standing position. He then realized that something similar could help anyone get around, especially on busy city streets.

How is the Segway HT different from other vehicles? It takes up less space and runs on rechargeable batteries instead of fuel. Like many of the machines that we use today, the Segway HT combines simple machines with high technology to help us do work in brand-new ways.

Glossary

compound machine (KAHM-pownd muh-SHEEN) a machine that combines two or more simple machines working together

counterweight (KOWN-tur-wayt) a weight used for balance on elevators and cranes

distance (DIS-tuns) in work, the space or area that an object travels as it moves

force (FORS) in work, something that sets an object in motion

fulcrum (FOOL-krum) the fixed point on which a lever turns

gear (GIR) a wheel with teeth. Two or more are used together.

inclined plane (in-KLYND PLAYN) a flat, slanted surface, such as a ramp

lever (LEV-ur) a bar or plane that turns on a fixed point

load (LOHD) the weight that's lifted or moved by a simple machine

pulley (POO-lee) a wheel with a rope looped around it

screw (SKROO) a thin pole or cylinder with a spiral ridge or thread wrapped around it

simple machine (SIM-pul muh-SHEEN) a simple device that helps you do work and has few or no moving parts

thread (THRED) the raised ridge on a screw

wedge (WEJ) an object with a pointed edge used to split or lift things

wheel and axle (HWEEL und AK-sul) a wheel that rotates around a rod

work (WURK) the act of moving something